ASTRONOSCOUT JOURNAL
SCIENTIFIC EDITION

THIS JOURNAL BELONGS TO

Astronoscout Journal is a product of Astronoscouts and Mythical Legends Publishing

ISBN: **978-1-943958-90-0**

CREDIT:

David Nash - http://www.astronexus.com/
An Introduction to Astronomical Terminology And Formulas licensed by a Creative Commons Attribution-ShareAlike license https://creative-commons.org/licenses/by-sa/2.5/

ASTRONOSCOUT JOURNAL
SCIENTIFIC EDITION

JANUARY 2019

Sun	Mon	Tues	Wed	Thu	Fri	Sat
		1	2	3	4	5
6	7	8	9	10	11	12
13	14	15	16	17	18	19
20	21	22	23	24	25	26
27	28	29	30	31		

FEBRUARY 2019

Sun	Mon	Tues	Wed	Thu	Fri	Sat
					1	2
3	4	5	6	7	8	9
10	11	12	13	14	15	16
17	18	19	20	21	22	23
24	25	26	27	28		

MARCH 2019

Sun	Mon	Tues	Wed	Thu	Fri	Sat
					1	2
3	4	5	6	7	8	9
10	11	12	13	14	15	16
17	18	19	20	21	22	23
24/31	25	26	27	28	29	30

APRIL 2019

Sun	Mon	Tues	Wed	Thu	Fri	Sat
	1	2	3	4	5	6
7	8	9	10	11	12	13
14	15	16	17	18	19	20
21	22	23	24	25	26	27
28	29	30				

MAY 2019

Sun	Mon	Tues	Wed	Thu	Fri	Sat
			1	2	3	4
5	6	7	8	9	10	11
12	13	14	15	16	17	18
19	20	21	22	23	24	25
26	27	28	29	30	31	

JUNE 2019

Sun	Mon	Tues	Wed	Thu	Fri	Sat
						1
2	3	4	5	6	7	8
9	10	11	12	13	14	15
16	17	18	19	20	21	22
23/30	24	25	26	27	28	29

JULY 2019

Sun	Mon	Tues	Wed	Thu	Fri	Sat
	1	2	3	4	5	6
7	8	9	10	11	12	13
14	15	16	17	18	19	20
21	22	23	24	25	26	27
28	29	30	31			

AUGUST 2019

Sun	Mon	Tues	Wed	Thu	Fri	Sat
				1	2	3
4	5	6	7	8	9	10
11	12	13	14	15	16	17
18	19	20	21	22	23	24
25	26	27	28	29	30	31

SEPTEMBER 2019

Sun	Mon	Tues	Wed	Thu	Fri	Sat
1	2	3	4	5	6	7
8	9	10	11	12	13	14
15	16	17	18	19	20	21
22	23	24	25	26	27	28
29	30					

OCTOBER 2019

Sun	Mon	Tues	Wed	Thu	Fri	Sat
		1	2	3	4	5
6	7	8	9	10	11	12
13	14	15	16	17	18	19
20	21	22	23	24	25	26
27	28	29	30	31		

NOVEMBER 2019

Sun	Mon	Tues	Wed	Thu	Fri	Sat
					1	2
3	4	5	6	7	8	9
10	11	12	13	14	15	16
17	18	19	20	21	22	23
24	25	26	27	28	29	30

DECEMBER 2019

Sun	Mon	Tues	Wed	Thu	Fri	Sat
1	2	3	4	5	6	7
8	9	10	11	12	13	14
15	16	17	18	19	20	21
22	23	24	25	26	27	28
29	30	31				

An Introduction to Astronomical Terminology And Formulas
by David Nash

My goal here is to make it easy for the amateur astronomer interested in stellar cartography (particularly 3D maps) and astrometry to do calculations. Unfortunately I can't avoid a lot of astronomical terminology and symbols, which might be a little confusing to the newcomer. If you're already comfortable with concepts such as *right ascension*, *proper motion*, *absolute magnitude*, and *parallax*, and know how to do calculations with them, you can safely skip this page. Otherwise, read on!

Right Ascension and Declination: Two Basic Coordinates

Just as locations on the spherical Earth can be specified with two numbers --latitude and longitude -- locations on the seemingly spherical sky can be given with a similar pair of numbers. There are many celestial coordinate systems -- this site uses one called *equatorial* coordinates. In effect, equatorial coordinates take the lines of latitude and longitude and expand them off the Earth into space. Thus the celestial analog of latitude measures how far "north" or "south" a star is from the Earth's equatorial plane, and the analog of longitude measures how far around the sky the star is.

The two coordinates have special names:

1. *Right ascension*, which is analogous to longitude, and
2. *Declination*, which is analogous to latitude.

In this web site, as in many other astronomical publications, the symbol for right ascension is α (alpha) and the symbol for declination is δ (delta).

For most practical purposes, you'll be looking up the right ascension and declination of stars on a star chart or in a list. Most of the time you don't have to worry much more about the details behind them -- just think of them as being like longitude and latitude and you'll be all right.

There are two exceptions, which are really important only if you plan on doing a lot of calculations:

1. For historical reasons α is normally expressed in *hours*, not degrees. One hour of right ascension equals 15 degrees, so right ascensions run from zero to 24 hours. Furthermore, hours of right ascension are often divided up into minutes and seconds, so you have to be careful in distinguishing seconds of angle (which come 3600 to a degree) from seconds of right ascension (which come 3600 to an hour).
2. α and δ are often listed in degrees (hours for α), minutes, and seconds in many catalogs. For astronomical calculations, these units are awkward -- it's much easier to work with "decimal" (ordinary floating point) values. Convert, if necessary, by:

    o α (decimal hours) = hours + minutes/60 + seconds/3600
    o δ (decimal degrees) = degrees + minutes/60 + seconds/3600

Remember, you may need to convert α (decimal hours) to degrees by multiplying by 15.

Parallax: Getting The Third Dimension

Since people can't perceive depth over the great distances of space, the stars appear to be imbedded in a flat, uniformly large sphere. Thus, for most purposes only two coordinates -- right ascension α and declination δ -- are needed to locate a star. However, astronomers have figured out that stars are at different distances from the sun, and so describing their positions fully requires a third coordinate. In most cases, the distance to the star serves as that third coordinate.

Although this is easy to understand mathematically -- three dimensions, three independent coordinates -- actually *measuring* the distance is extremely difficult. Astronomers did not succeed until the 1830s -- more than two centuries after the invention of the telescope -- when they finally succeeded in measuring a quantity called *parallax*. Parallax is the result of motion -- as seen from a moving object, nearby objects appear to shift with respect to farther ones. Since the Earth revolves around the Sun once per year, nearby stars show a small but measurable shift during that time.

Dealing with parallax

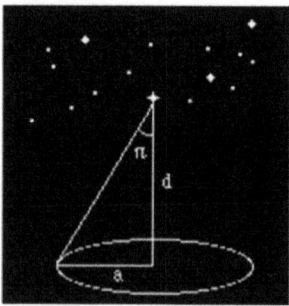

Parallax is expressed as an angle, and given the symbol π (totally unrelated to the symbol for 3.14159...). It depends on the radius of the Earth's orbit (a) and the star's distance (d). The usual formula for parallax is:

- $\tan \pi = a / d$

However, since it's true for all stars that d >> a, you can use a small-angle approximation (tan x ~ x) and write:

- $\pi = a / d$

This formula gives π in *radians*. Radians are awkward, though -- the parallax angles are always much less than 1 radian, and so astronomers prefer to measure parallax in arcseconds (1/3600 degree, or roughly 1/200,000 radian). Astronomers also define a distance unit from parallax data, one specially chosen to be convenient for working with parallaxes. The *parsec*, abbreviated pc, is defined as the distance d at which the parallax π equals exactly 1 arcsecond. If the distance to a star is given in parsecs, and the parallax is given in arcseconds, then:

- $\pi = 1 / d$

To get the distance to a star in parsecs, just take the reciprocal of the parallax. Therefore, parsecs are a very convenient unit when using parallax data, and so I use parsecs rather than light years throughout this site. If you prefer working with light years, 1 pc = 3.262 light years.

The closest bright star to the sun, Alpha Centauri, has a parallax of 0.742 arcseconds and a distance of 1/0.742 = 1.35 pc. Only about 60 stars have distances under 5 pc, and many of the most famous bright stars, such as Betelgeuse in Orion, and Polaris the North Star, are many hundreds of parsecs away.

Magnitudes: Brightnesses Apparent and Actual

In addition to knowing *where* the stars are, it's important to know how bright they are. Astronomers use a system called *magnitude* to characterize the brightness of a star.

Apparent Magnitude

Apparent magnitude, V, is the brightness of the star as it appears from Earth. Most star catalogs give the apparent magnitude of a star along with its position. Larger magnitudes denote *fainter* stars -- this seems a little weird initially, but if you think of "first class" versus "third rate" it will make more sense.

The magnitude scale is logarithmic, with a brightness ratio of 100 corresponding to a difference of 5 magnitude units. Naked eye stars typically have magnitudes from about 0 to +6; Sirius, the brightest star visible in the night sky, has a magnitude of -1.44. By contrast, stars dimmer than magnitude +10 can be seen only in binoculars or telescopes, and those dimmer than +20 require giant telescopes to be seen at all.

Absolute Magnitude

Since stars are at very different distances, stars of similar apparent magnitudes may be very different in total light output. Astronomers need a more fundamental measure of stellar energy output. The *absolute magnitude* M_V is defined as the apparent magnitude (V) a star would have if it were moved to a distance of 10 parsecs.

If a star's apparent magnitude V and distance d are known, you can find M_V from the following formula:

- $M_V = V + 5 \log_{10} (10 / d)$

As an example, Sirius has an apparent magnitude of -1.44, and its distance is 2.64 parsecs. Thus the absolute magnitude of Sirius is:

- $M_V = -1.44 + 5 \log_{10}(10 / 2.64)$
- $M_V = +1.45$

Just as a 100-fold difference in apparent brightness yields a 5 unit change in apparent magnitude, a 100-fold difference in *true luminosity*, i.e., actual energy output, yields a 5 unit difference in absolute magnitude. Thus comparing two stars' absolute magnitudes shows which star is more luminous, and by how much.

The Sun's absolute magnitude is +4.85. The least luminous stars -- small cool stars with low masses -- have absolute magnitudes of about +18, whereas supernovas, stars undergoing colossal explosions, can have absolute magnitudes of -15 or brighter.

Stellar Motions

Although stars appear stationary, with the constellations keeping their shape and appearance, the stars are actually moving around rapidly. Only the great distances between stars keep us from noticing this readily. However, with modern equipment, astronomers can easily measure these motions.

Proper Motion

A star that is moving through space will normally appear to move across our line of sight to it -- that is, the star will appear to move sideways and hence shift with respect to other stars. This motion is called *proper motion*, because it is an intrinsic motion of the star, rather than an apparent motion caused by the motion of the Earth. Proper motion, denoted by μ, is expressed as an angular velocity (usually in arcseconds per year).

A star's proper motion can be in any direction. Usually, proper motion is decomposed into two components. One, (μ_α) corresponds to change in right ascension, and one (μ_δ) to change in declination. These two components are what you'll normally find in catalogs.

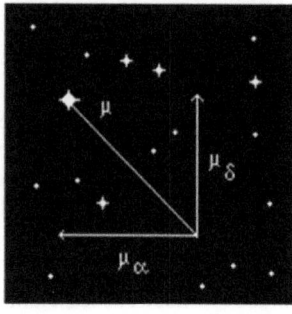

In calculus terms, $\mu_\alpha = d\,\alpha\,/\,dt$ and $\mu_\delta = d\,\delta\,/\,dt$ (in appropriate units, of course).

Transverse Velocity

Proper motion is an angular velocity. If the distance to a star is known, the proper motion can be turned into a *linear* velocity -- one with units like kilometers per second. The linear velocity corresponding to proper motion is *transverse velocity*, V_T. Transverse velocity is important for calculating stellar motions over time -- a calculation discussed in more detail in the Long-Term Stellar Motions page.

Radial Velocity

Stars can also move along our line of sight to them. The component of a star's motion towards or away from Earth is the *radial velocity*, V_R. Unlike proper motion, radial velocity is a linear velocity, and has units such as kilometers per second. Radial velocity can be measured accurately by analyzing a star's light in detail.

Total Space Velocity

Finally, the star's *total* velocity, through space, is the vector sum of its radial and transverse velocities. The relationship between the star's total velocity, v, and its radial and transverse velocities v_R and v_T, is illustrated in the following diagram:

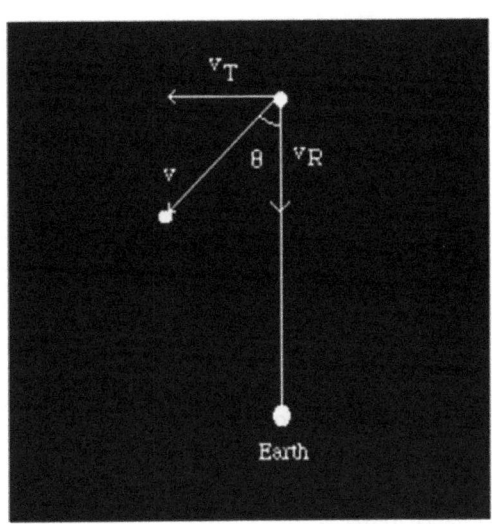

Common Astronomical Symbols

Symbol	Meaning
Equatorial Coordinates	
α	Right ascension
δ	Declination
d	Distance to star
General Spherical Coordinates	
θ	Azimuth (spherical coordinates)
ϕ	Altitude (spherical coordinates)
r	Distance to star
Angular Velocities	
μ_α	Component of proper motion in right ascension
μ_δ	Component of proper motion in declination
Linear Velocities	
v_R	Radial velocity (negative = approaching)
v_{TA}	Transverse velocity, component in right ascension
v_{TD}	Transverse velocity, component in declination
v_x, v_y, v_z	Cartesian velocity vectors
Other Quantities	
t	Time (usually in years)
π	Parallax
V	Apparent visual magnitude
M_V	Absolute visual magnitude